WHAT'S SO SCARY ABOUT

BATS

Joanne Mattern

RED CHAIR PRESS

What's So Scary is produced and published by Red Chair Press:
Red Chair Press LLC PO Box 333 South Egremont, MA 01258-0333
www.redchairpress.com

FREE Educator Guides at www.redchairpress.com/free-resources

Publisher's Cataloging-In-Publication Data
Names: Mattern, Joanne, 1963- author. | Mattern, Joanne, 1963- Earth's amazing animals.
Title: What's so scary about bats? / Joanne Mattern.

Description: South Egremont, MA : Red Chair Press, [2022] | Series: Core content science : Earth's amazing animals | Interest age level: 008-010. | Includes index and suggested resources for further reading. | Summary: "Bats have a bad reputation as blood-sucking creatures of the night. But these fascinating mammals actually help people by eating insect pests and pollinating fruit trees and flowers. In this book packed with fun facts and vivid photographs, readers will learn all about the bat's amazing body, its habits and activities, and how they can help this important species survive"--Provided by publisher.

Identifiers: ISBN 9781643711645 (hardcover) | ISBN 9781643711683 (softcover) | ISBN 9781643711720 (ePDF) | ISBN 9781643711768 (ePub 3 S&L) | ISBN 9781643711805 (ePub 3 TR) | ISBN 9781643711843 (Kindle)

Subjects: LCSH: Bats--Juvenile literature. | CYAC: Bats.

Classification: LCC QL737.C5 M38 2022 (print) | LCC QL737.C5 (ebook) | DDC 599.4--dc23

Library of Congress Control Number: 2021944658

Photo credits: Cover, p. 1, 3, 5, 8, 11, 12, 13, 15, 17, 19, 21, 24, 25, 28, 29, 30, 31, back cover: iStock; p. 7, 9, 14, 16, 18, 20, 23, 27, 29: Shutterstock

Printed in United States of America
0422 1P CGF22

Table of Contents

Scary Bats!

A dark shape swoops out of the night sky. The creature flutters its wings. Sharp teeth are ready to bite. The animal dives toward you. You scream in fear. It's a bat!

Many people are scared of bats. Some people think bats drink blood. They may even believe bats can turn into vampires. Other people just don't like the way bats swoop and fly in the night sky. Are bats creepy? Maybe. But they are also very helpful. Let's learn about these creatures of the night and why bats are really not so scary after all.

The fruit bat has wings made of a thin layer of skin.

Bat Basics

Bats are mammals. Like other mammals, bats are warm-blooded. They are covered with fur. Mammals give birth to live young. They nurse their babies with milk from their own bodies.

There are two kinds of bats. Microbats are small. Most of these little bats eat insects. The second kind of bat is called megabats. Megabats are bigger. They eat fruit.

Bats can live in almost any habitat. You'll find them in woods, deserts, and cities. The only continent that doesn't have bats is Antarctica. It is just too cold there for bats to survive.

Now You Know!

Kitti's hog-nosed bat is the smallest bat in the world. It is smaller than a person's thumb! This bat is also called the bumblebee bat, because it is about the same size as a bumblebee. These tiny bats live in Thailand and Myanmar in Southeast Asia.

Microbats are usually under 6 inches (16 cm) long.

Bat Bodies

A microbat's body is small and furry. These bats have big ears that face forward. They also have a short tail and tiny feet.

A microbat's body is small, but its wings are big! A bat's wings are made of thin skin stretched over four fingers and a thumb. A thin **membrane** connects a bat's wings to its body.

Megabats also have big wings. The wings can spread out more than three feet (1 meter). They use their fingers to hold fruit while they eat. Megabats have long noses and big eyes.

Now You Know!

Megabats are also called fruit bats or flying foxes.

Insect Eaters

Microbats are **predators**. Most microbats eat insects. How do they find insects in the dark night? The answer is **echolocation**.

A bat makes high-pitched squeaks as it flies. These squeaks bounce off objects and back to the bat's big ears. The bat listens to the echoes and uses them to identify insects in the air. Then the bat swoops down and gobbles up its **prey**.

One bat can eat hundreds of mosquitoes in one night. They also eat moths and beetles. Bats are great helpers at pest control!

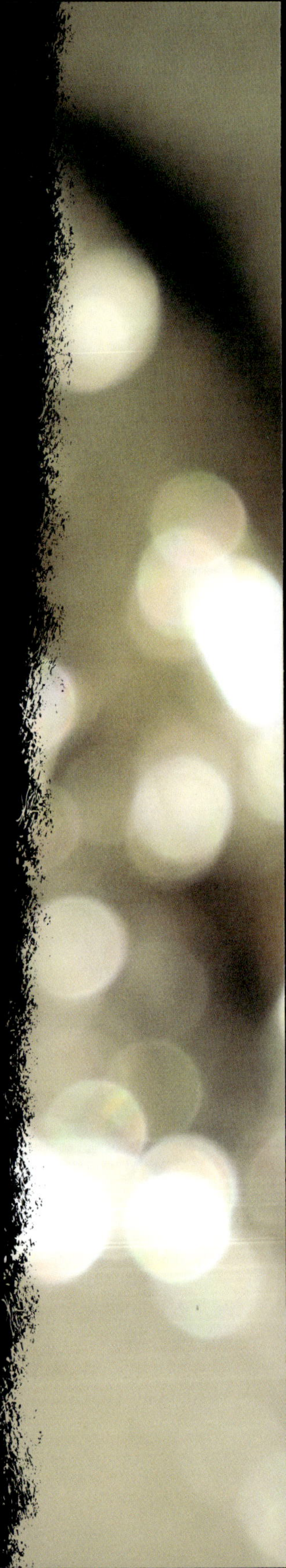

Now You Know!

The fish-eating Myotis bat drags its long legs and clawed feet through the water to grab any fish in its path.

This microbat in Kenya is feeding on a locust.

We Love Fruit!

Megabats don't eat insects. They like fruit instead. These bats live in hot, wet places in Australia, Africa, and Asia. They use their senses of smell and sight to find tasty fruit to eat. Avocadoes, peaches, mangoes, and dates are some of the fruits megabats find tasty.

Megabats help plants in many ways. When they land on a plant, some **pollen** sticks to the bat's body. Later, when the bat lands on another plant, that pollen falls off. This helps plants **reproduce**.

Megabats also swallow seeds as they eat. Later, they poop out those seeds. The seeds fall on the ground and grow new plants.

Home Sweet Bat Home

Bats live in many different places. Some bats live in caves. These places are dark and safe. Thousands of bats **roost** in a cave. They sleep hanging upside-down during the day. At night they fly out to hunt.

Many bats live in cities. They roost under bridges. They sleep in attics and other spaces inside buildings.

Fruit bats roost in trees. They stay safe under the branches.

Now You Know!

Every night, people gather to watch thousands of Mexican free-tailed bats fly out from under the Congress Avenue Bridge in Austin, Texas.

EDITOR'S NOTE: Never touch a bat with your bare hands.

Baby Bats

In early summer, female bats find a safe place to have their babies. These places are called maternity roosts. Some bats return to the same maternity roosts every year.

A bat gives birth to one or two babies at a time. Baby bats are called pups. They are tiny and helpless. They stay close to their mother for about five weeks. During this time, they feed on their mother's milk.

After four or five weeks, the pups are able to fly. They make short trips out of the roost to catch prey. Soon the little bats are able to hunt for themselves. It's time to leave Mom and go out on their own.

Indian Flying Fox with her pup under her wing.

Meet the Bats

Here are a few of the many **species** of bats on Planet Earth:

Little Brown Bat

This microbat lives in the United States and Canada. It catches lots of mosquitoes and other insects. It uses echolocation to find its prey.

Little brown bat

Pipistrelle bat

Pipistrelle Bat

This bat is found all over Europe, Africa, and Asia. It lives in the woods and near water. Pipistrelles like to eat mosquitoes, gnats, flies, and other insect pests.

Golden-crowned Flying Fox

This megabat is really mega! It is one of the largest bats in the world. Its wingspan is more than five feet (1.5 meters) wide, and it weighs about two and a half pounds (1 kilogram). This bat lives in the Philippines and eats figs and other fruit.

Grey-Headed Flying Fox

This megabat lives along the eastern and southern coasts of Australia. It's easy to spot because of the orange fur around its neck. Thousands of these bats can live together in the forests. They fly along waterways to find food.

Grey-headed flying fox

Are Bats Vampires?

Many old horror movies show vampires changing into bats, or bats changing into vampires. A bat even looks like a vampire when it spreads out its wings like a cape. Like vampires, bats are **nocturnal**. The idea of a bat changing into a monster and drinking your blood is very scary. But it is mostly movie magic.

However, there is one species of bat called the vampire bat. And yes, this bat does drink blood. The vampire bat lives in South and Central America. It goes out at night and finds a sleeping mammal, such as a cow or a sheep. The bat uses its sharp teeth to make a small cut. Then it laps up the blood from the wound.

Vampire bats are kind of scary. But they rarely bite people, and they won't suck your blood.

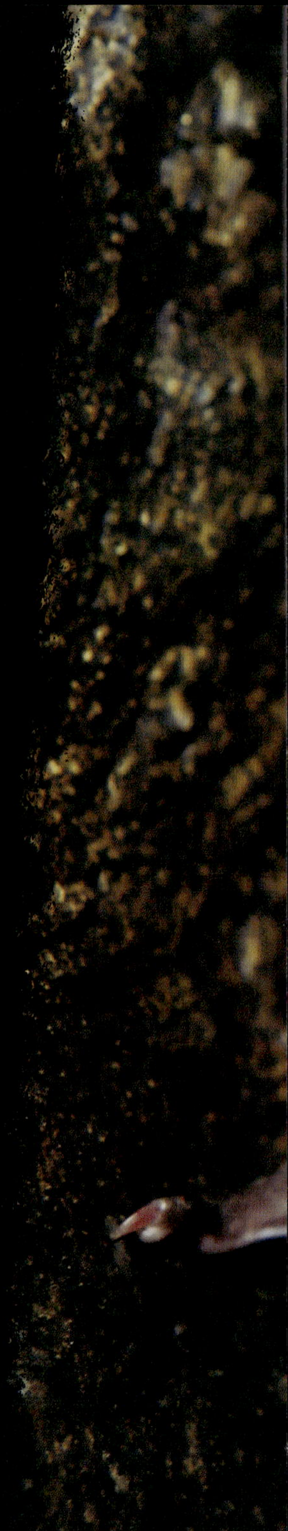

Vampire bat

Now You Know!

One type of vampire bat sucks blood from birds.

Warning! Danger Ahead!

Bats are amazing creatures that help people in many ways. They eat lots of harmful insects, spread seeds, and help plants reproduce. Bats are not truly scary. But they are wild animals, and they can be dangerous.

Bats can spread a scary disease called **rabies**. By the time people realize they have rabies, it may be too late to save them. Many animals, including dogs and raccoons, spread rabies by biting. Bats do too. Bats can also spread rabies through a scratch. Even touching bat saliva can cause a person to catch rabies. Anyone who has touched a bat or found one in their home should check with a doctor to be safe.

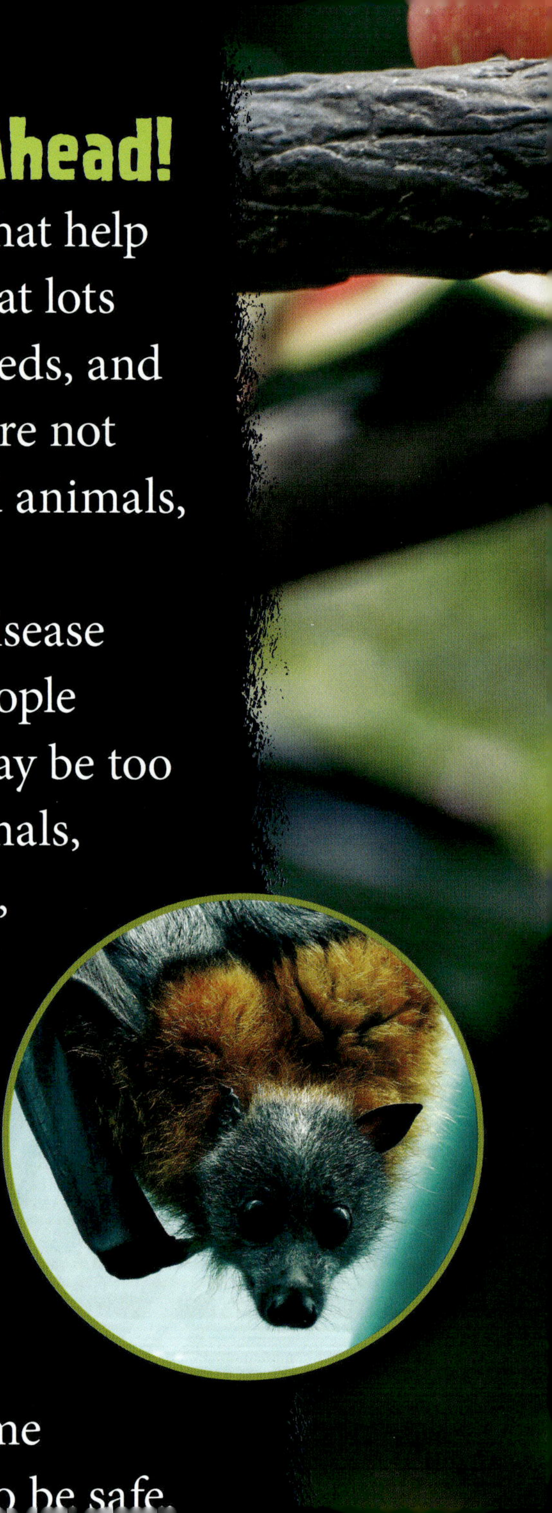

Face-to-Face With a Bat

Sometimes bats get into people's houses. These bats might live in the attic or another part of the house. Or they might fly in from outside. A bat in a house will often stay near a light on the ceiling because insects are also attracted to that light.

What should you do if there is a bat in your house? Do not try to catch the bat with your bare hands. Wear thick gloves and long sleeves so you don't get scratched or bitten. Try to shoo the bat outside using a broom. If you can't get rid of the bat, or if there are more than one, shut the door to the room. Then call a pest control company, or animal control in your town, to help. They have the right tools and know how to do the job safely.

Now You Know!

Many people are afraid a bat will get tangled in their hair. However, most bats don't want to get close to a person. And getting tangled in hair doesn't sound like much fun for the bat!

Bats in Trouble

Here's a scary fact: Bats are in big trouble. Since 2006, millions of bats have died from a disease called white-nose syndrome. This disease comes from a **fungus** that grows on bats while they hibernate, or sleep through the winter. Since large groups of bats hibernate close together, the fungus spreads quickly. It makes the bats so thirsty that they wake up. Because it is cold and there is nothing to eat in the winter, the bats waste a lot of energy trying to stay warm. In a short time, the bat starves to death.

Scientists are working hard to find a cure for white-nose syndrome. They want to save as many bats as they can because of all the ways they help us.

Wildlife biologists check these brown bats for white-nose syndrome.

Helping Bats

Many people are trying to help bats. They know that without bats, we would have a lot more insect pests. Bats are a natural way to get rid of harmful insects.

One way to help bats is to build a bat house. These homes provide a safe place for bats to roost and live.

Another way to help bats is to give them space. Many caves and old mines are gated off. This means people can't get in and bother the bats living there. Keeping bats safe makes our world a better, not-so-scary place!

Glossary

echolocation finding objects by using reflected sound

fungus a type of plant that reproduces by spores

membrane a thin layer of skin

nocturnal active at night

pollen tiny grains produced by flowers that help them reproduce

predators animals that hunt other animals for food

prey animals that are hunted by other animals for food

rabies a deadly disease that attacks the brain and is spread by animal bites

reproduce to produce young

roost to sleep in a safe place

species a group of living things of the same kind

Learn More in the Library

Baxter, Roberta. *Native Pollinators: Bats*. Mitchell Lane Publishers, 2020.

Klatte, Kathleen A. *Bats at Night*. PowerKids Press, 2021.

Laidlaw, Rob. *Bat Citizens: Defending the Ninjas of the Night*. Pajama Press, Inc., 2018.

Mattern, Joanne. *Fantastic Homes for Flying Creatures*. Red Chair Press, 2023.

Milner, Charlotte. *The Bat Book*. DK Publishing, 2020.

Index

About the Author

Joanne Mattern is the author of many books for children including the previous three sets of *Core Content: Earth's Amazing Animals Series.* She loves writing about sports, all kinds of animals, and interesting people. Mattern lives in New York State with her family.